YOUR KNOWLEDGE HAS VALUE

- We will publish your bachelor's and
  master's thesis, essays and papers

- Your own eBook and book -
  sold worldwide in all relevant shops

- Earn money with each sale

Upload your text at www.GRIN.com
and publish for free

Bibliographic information published by the German National Library:

The German National Library lists this publication in the National Bibliography; detailed bibliographic data are available on the Internet at http://dnb.dnb.de .

Imprint:

Copyright © 2018 GRIN Verlag
Print and binding: Books on Demand GmbH, Norderstedt Germany
ISBN: 9783668809529

This book at GRIN:

https://www.grin.com/document/442593

Manoj Kandel, Surya Kant Ghimire, Bishnu Raj Ojha, Jiban Shrestha

Evaluation of maize (Zea mays L.) inbred lines under heat stress and normal condition

GRIN Verlag

Evaluation of maize (*Zea mays* L.) inbred lines under heat stress and normal condition

Table of contents

Abstract 2

Introduction 3

Materials and method 4

  Data collection 5

  Statistical analysis 6

    Genotypic and phenotypic coefficient of variation 6

    Heritability (h^2) 6

    Genetic advance 6

    Stress susceptible index (SSI) 7

    Stress tolerance index (STI) 7

    Tolerance index (TOL) 7

    Geometric means productivity (GMP 7

    Mean productivity (MP) 7

  Results and Discussion 7

    Anthesis-silking interval (ASI) 8

    Leaf firing and Tassel blast 8

    SPAD chlorophyll and leaf senescence 8

    Plant and Ear Height 9

    Leaf area index (LAI) 9

    Ear Per Plant (EPP) 9

    Number of kernel per ear (NKE^{-1}) 9

    Silk Receptivity 10

    Shelling percentage 10

    Thousand Kernel weight (TKW) 10

    Grain Yield 11

    Coefficient of Phenotypic and Genotypic Variation (PCV and GCV) 15

    Heritability (H^2b) and Genetic advance 15

    Correlation analysis 16

    Path coefficient analysis 19

    Stress tolerance indices 22

    Principal Component analysis 29

Conclusion 32

Acknowledgments 33

References 34

Evaluation of maize (*Zea mays* L.) inbred lines under heat stress and normal condition

Manoj Kandel[1], Surya Kant Ghimire[1], Bishnu Raj Ojha[1] and Jiban Shrestha[2]

[1]Department of Genetics and Plant breeding, Agriculture and Forestry University (AFU), Rampur, Chitwan, Nepal
[2]Nepal Agricultural Research Council, National Maize Research Program, Rampur, Chitwan, Nepal

Abstract

Twenty maize genotypes were tested in alpha-lattice design with two replications in research block of National Maize Research Program, Rampur, Chitwan from February to June, 2016 to evaluate maize inbred lines for their genetic components association among yield components and correlation and stress tolerance indices. Meteorological data showed maximum mean temperature (46.2–43.28°C) and minimum (30.52-30.77^{0}C) in with relative humidity 37.05 to 49.45% inside the tunnel during in April-May which coincide with the flowering and grain filling periods. Mean temperature of during the reproductive stage of plastic house was 8-9°C higher than normal condition. Six selection indices for stress tolerance including heat susceptible index (HSI),stress tolerance index (STI), susceptible index (SI), tolerance index (TOL),geometric means productivity (GMP), mean productivity (MP) were calculated based on grain yield of maize under heat stress and normal conditions. Secondary traits such as anthesis–silking interval (ASI),leaf firing, tassel blast, SPAD reading and leaf senescence, plant and ear height, leaf area index, ear per plant, cob length and diameter, number of kernel ear^{-1}, number of kernel row^{-1} and number kernel ear^{-1},silk receptivity, shelling percentage, thousand kernel weight were recorded. Shorter ASI was found in RML-17,RML-57, RL-107, RL-140, RML-91 whereas more leaf area index in NML-2, RML-91, and RML-24 respectively. Inbred lines, RML-91,RML-17, RML-20, RL-140, RL-140, RML-91 and NML-2 were found lower tassel blast and leaf firing percentage respectively. Inbred lines, RML-76,RML-115,RML-4,RL-111 showed slower leaf senescence and RML-57,RL-101,RML-11,RL-107 showed significant higher value of SPAD chlorophyll content.RML-91,RL-111,NML-2 showed higher plant and ear height.RML-40,RML-91,RML-7 and RML-91,RML-20,RL-107 showed higher cob diameter and length respectively. Inbred lines RML-91,RML-4,RML-115 exhibited longer physiological maturity and RL-105,RML-91,RML-20 showed more ear per plant. Inbred lines RML-91,RML-7,RL-101,RL-140,RML-76,RML-91and RML-91,RL-140,RML-76 was associated with higher number of kernel ear^{-1}, number of kernel row^{-1} and number kernel ear^{-1} respectively. RML-91,RL-140,RML-7 and RML-17,RL-101,RML-40 showed maximum silk receptivity and shelling percentage.RML-7,RML-17,RL-140 and RML-91,RL-140,RML-7,RML-4 showed higher thousand kernel weight Zend grain yield respectively. Heritability and correlation analysis revealed that grain yield, number of kernel ear^{-1},silk receptivity, shelling percentage, thousand kernel weight plant and ear heights showed presence of additive gene effect and those traits direct could be used as direct selection to improve maize grain yield under heat stress condition. Anthesis silking interval, leaf firing, tassel blast, leaf senescence, ear per plant, leaf area index, cob length, number of kernel row ear^{-1}showed non-additive or dominant gene action and suggested as hybrid plants have higher capacity to tolerate heat stress in normal conditions than their parents. With attention to high negative correlation with anthesis silking interval, tassel blast, leaf firing with grain yield the minimum of anthesis silking interval, tassel blast and leaf firing were the most important index required further genotyping for development of thermo-tolerance lines. Cluster and PCA analysis revealed that cluster 4 genotype named as RL-140,RML-76,RML-40 and RML-91 were found to be tolerant to heat stress with higher value of grain yield and other desirable traits and lower leaf firing, Tassel blast and itermediate of anthesis-silking interval which were suitable for cultivation under heat stress condition.Based on stress tolerance indices and their correlation analysis, RML-91,RML-140 appeared as having high yield potential and low stress susceptibility under normal and heat stress condition.

Keywords: Maize, Heat stress, Morpho-physiological traits, Correlation, Cluster, Heat indices

Introduction

Maize being nutritionally an important crop has multiple functions in thetraditional farming system; being used as food andfuel for human beings andfeed for livestock andpoultry. In Nepal, it is grown in 8,91,583 ha producing 2.2 million tons, with an average yield of 2500 kg ha^{-1} (MOAD, 2016).The reasons for low maize yield in Nepal are high temperature, drought, stalk rot infestation, maize borer and shoot fly infestation, poor crop management, high input rates and use of low quality, substandard seed.

Heat and drought stress have emerged as a common problem worldwide which can reduce maize crop productivity (Ali *et al.*, 2015). Heat stress in the flowering and grain filling periods due to elevated temperatures drastically affect crop productivity. A record drop in maize production was reported in many maize-growing areas of the world (Van der Velde *et al.*, 2010).It is predicted that maize yield might be reduced up 70 % due to increasing temperatures (Khodarahmpour *et al.*, 2011).A report of the Asian Development Bank (2009) warns that if the current trends persist until 2050 major food crop yields and food production capacity of south Asia will significantly decreases by 17% for maize,12% for wheat and 10% for rice due to climate change induced heat and water stress.

Maize crop yield potential grown in terai is always at risk from important biotic and abiotic stresses which limit crop production. Now day's heat stress is one of key abiotic stress with high potential impact on maize crop growth and development and eventually on productivity. Various plant organs, in a definite hierarchy and in interaction with each other are involved in determining crop yield under stress (Barnabas *et al.*, 2008). The response of maize crop to climate depended on the genetic and physiological make up of variety being grown and interaction with prevailing climatic condition. Therefore in comparison to agronomic management genetic management of heat stress tolerance genotypes would be low economic input technology that would be readily acceptable to resource's –poor, heat affected and small land holding farmer (Saxena & Toole, 2002).

Transitory or constantly high temperatures cause an array of morph-physiological, anatomical and biochemical changes in plants, which eventually affects plant growth and development, and lead to a drastic reduction in biological and economic yield (Commuri & Jones, 2001).The correlation studies measure the associationsbetween yield and other traits. Path coefficientanalysis permits the separation of correlationcoefficient into direct and indirect effects. Therefore, thepresent investigation was carried out to determinethe association of traits with grain yield throughcorrelation coefficient and direct and indirecteffect of a set of variables through path analysisunder heat stress condition in maize.

Materials and method

The research was conducted at National Maize Research Program (NMRP) of Rampur, Chitwan during spring season from February 24, 2015 to July 2016, geographically located at 27° 37' North Latitude and 84 ° 29' East longitude at an altitude of 225 meter above sea level. This site contains only sandy loam soil with acidic reaction. This research location is characteristics of subtropical climate. The plant materials were collected from National Maize Research Program (NMRP).The list of inbred lines along with pedigree information included in the study is presented in Table 1.

Table 1: Names and pedigree information of maize inbred lines used for heat stress research at NMRP Chitwan (2016).

S.N	Inbred line of maize	Pedigree sources/origin	S.N	Inbred line of maize	Pedigree sources/origin
1	NML-2	CML-430	11	RL–101	UPAHAR-B-20-2-3-1-1
2	RML–4	CA00326	12	RML–24	CA00304
3	RML–32	CA00320	13	RML–40	CML-427
4	RML–95	PUTU-17	14	RML–57	CLQG6602
5	RML–86	PUTU-20	15	RL–107	UPAHAR-B-20-2-4-3-1
6	RML–17	CML-287	16	RML–20	CA-34503
7	RML–96	AG-27	17	RML–76	CLRCYQ007
8	RL–105	UPAHAR-B-20-2-4-1-1	18	RML–7	CML-413
9	RL–111	UPAHAR-B-31-1-1-1-1	19	RML–91	PUTU-19
10	RML–115	PUTU-17	20	RL–140	POOL-21-12-1-2-2-1-1

Field experiment was conducted in alpha lattice design. There were two conditions: normal and plastic house (for heat stress), each condition replicated twice. Each replication comprised four blocks consisting of five plots each. Each plot was 3 meter in length 0.6 meter wide. Each plot had one row with spacing 20 cm between rows, inter block gap was 0.5 m was maintained. Each plot contained single row with spacing 60×20 cm and consisted 15 hills, each of two seed were sown, one of whose seedling were removed at the six leaves stage. The dose of chemical fertilizer applied was 120:60:40 kg NPK per hectare. Fertilizer were applied prior to sowing at rate of 60 N kg ha^{-1}, 60 kg P and 40 K ha^{-1} and additional side dressing of 30 N kg ha^{-1} were applied at the two times in six leaves stage and knee high stage of maize. The irrigation was done three important stage, knee high stage, tasseling stage and milking stage. To created heat stress condition maize study half of field was controlled heating imposed using two plastic (120gsm) houses were used two week just prior to the onset of reproductive period up to the crop harvesting. Maximum mean temperature 46.2°C in April in heat stress condition whereas as normal condition was 37.23°C and similarly for May month in maximum mean temperature was 43.28°C whereas in normal condition 34.54°Cat time of flowering, pollination and grain filling periods as shown in Table 2.Partial opening top side of tunnel was done for control relative humidity inside tunnel to avoid any possible disease outbreak.

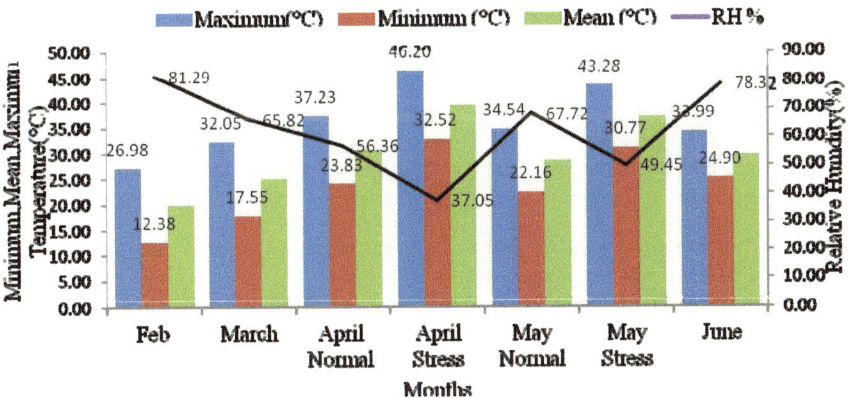

Figure 1 :Weather data recorded in NMRP, Chitwan during Experimental period (24th Feb to June 2016 Source: National Maize Research Program (NMRP), NARC, Chitwan

Data collection

Data on days to 50% anthesis, days to 50% silking and ears per plot, leaf firing, tassel blast, leaf senescence were recorded on plot basis. Whereas, ear height (cm), number of kernel ear^{-1}, number of kernel row, number of kernel row^{-1},SPAD reading, leaf area index, silk receptivity, thousand kernel weight (g) and shelling per cent were recorded on five selected representative plants. The sample cobs were shelled, cleaned and grain weight and shank weight were recorded to calculate the shelling percent. Thousand kernel weight was measured by counting 1000 grains from the bulk of each plot after shelling and weighed in grams after the moisture was adjusted to 15%. Anthesis-silking interval (ASI) was calculated by subtracting the number of days taken for 50% anthesis from the number of days taken to50% silk emergence. Leaf area index was calculated by total leaf area divided by land area and multiply by correction factor (0.75). Silk receptivity was recorded by total number of fertilized grains per ear divided by number of potential grain per ear. Leaf firing was obtained by the counting the number of plants that showed leaf firing symptoms in the total number of plants in a particular plot and was expressed in percentage. Tassel blast was obtained by the counting the number of plants that showed tassel blast symptoms in the total number of plants in particular plot was expressed in percentage. Grain mass per plot adjusted to 15 % grain moisture and converted to kilogram per hectare on basis by using following formula

$$\text{Grain yield (Kg ha}^{-1}) = \frac{\text{FEW} \times \text{SP} \times (100 - \text{GMC})}{\text{NHA} \times 85 \times 10}$$

Where,

FEW = field ears weight (Kg), SP = shelling percentage (%), GMC = grain moisture content at harvest (%), NHA = net harvested area (m^2)

Statistical analysis

The data recorded on different parameters from in heat stress field were first tabulated and processing in Microsoft excel (MS- Excel, 2010), then subjected to restricted maximum likelihood (REML) tool in GenStat to obtain ANOVA. Correlation coefficients of different traits were carried out using the formula given by Steel & Torrie (1980) by using SPSS program. Path analysis carried out by using MS- Excel. The biplot display was used, which provides a useful tool for data analysis. To display the genotypes in biplot, a principal component analysis was performed by using Minitab 17. Coefficient of variation

It is defined as ratio of standard deviation to mean expressed in percentage. More value of C.V. indicates less will be homogeneity.

Genotypic and phenotypic coefficient of variation

The genotypic and phenotypic coefficient was calculated according to Burton & Devane, (1953) and expressed as a percentage.

Phenotypic coefficient of variation (PCV) = (Phenotypic SD/General mean of character) × 100

Genotypic Coefficient of variation (GCV) = (Genotypic SD/General mean of character) ×100
PCV and GCV value were categorized by Sivasubramanian and Menon (1973) as given below:

0-10 % = Low, 10-20% = Medium,>20= High

Heritability (h^2)

Broad sense heritability (h^2_b) for the trait was estimated by using the variance components method suggested by Becker, (1984), as follows:

$$H^2b = \frac{\text{MSS due to genotype} - \text{MSS due to error}}{\text{Replication}}$$

Where, h^2b= broad sense heritability, r= number of replication

Genetic advance

The extent of genetic advance to be expected by selecting five percent of the superior progeny was calculated by using the following formula given by Robinson *et al.* (1949).

$GA = I\ \sigma_p H^2_b$

Where, I = efficiency of selection which is 2.06 at 5% selection intensity, σ_p= phenotypic standard deviation,H^2_b= heritability in broad sense. Different indices were carried out using the given formula as given below:

Stress susceptible index (SSI)
Stress susceptible indices for each genotype were calculated as
$$SSI= [\frac{1-(YS \div Yp)}{D}]$$
Where,
Y_P = Yield of genotype under non-stress condition
Y_S= Yield of Genotype under stress condition
D= Stress intensity = $1-\frac{Meand\ of\ XP}{Means\ of\ XS}$
XP= mean of all genotype yield under normal condition
XS= Means of all genotype yield under stress condition
Genotype was categorized as tolerant and susceptible. Genotype having HIS≤0.50 were highly tolerant, HIS>0.50<1.0 were moderately tolerant and HIS>1.0 were susceptible.

Stress tolerance index (STI)
It was given by Fernandez (1992).
$$STI= \frac{YP \times YS}{(YP)^2}$$
Where the symbols have similar meaning as given earlier. High value STI indicates the genotype to be more tolerant to the one with lower value. Selection based on STI resulted in selection of genotype with higher stress tolerance and yield potential.

Tolerance index (TOL)
It was given by Rosielle and Hamblin (1981) as follows
$$TOL= Y_P - Y_S$$

Geometric means productivity (GMP)
It was calculated as given by Fernandez (1992) as follows
$$GMP= \sqrt{YP} \times YS$$
For this genotype with higher value were selected as higher indicate more tolerant genotype.

Mean productivity (MP)
It is average of yield under stress and normal environment. It was given by Rosielle and Hamblin (1981). Selection of genotype based on mean productivity was done for those having values which indicate them to be tolerant. $MP= \frac{YP+YS}{2}$

Results and Discussion

Maximum mean temperature 46.2°C in April in heat stress condition whereas as normal condition was 37.23°C and similarly for May month in maximum mean temperature was

43.28°C whereas in normal condition 34.54°Cmeans mean temperature 8-9°C higher in plastic house at time of flowering, pollination and grain filling periods as well as there was no drought due to plot was irrigated when the surface began dry and all other factor were kept constant, the only stress led to difference between the normal and stress treatment was assumed to be due to heat stress.Analysis of variance was performed for all traits. Analysis of variance of mean square comparison described significant difference among 20 maize inbred lines for traits as shown in Table 2 and 3.

Anthesis-silking interval (ASI)

In present study, significant increases the ASI under heat stress as compared to normal condition was observed. Mhike *et al.*(2012) also reported similar result significant increases the ASI in maize under heat stress might be silking delay i.e due to at rise in temperature (+30°C), pollen shedding starts much ahead of silks emergence while silking is delayed, so that silking period does not correspond to anthesis/tasseling, resulting in poor synchronization of flowering.Cicchino *et al.*(2010) reported mechanisms of increases anthesis silking interval under heat stress condition causes reduction in essential nutrients as well as all other factors like pollen viability, increase silking day known as silk delay.Edmeades *et al.* (1993) reported association of short anthesis-silking interval with increased partitioning of assimilate to the ear at flowering and supply of nitrogen to the developing ear a cause of improved performance of genotypes under stress environment.

Leaf firing and Tassel blast

Leaf firing of the flag leaf and one or two adjacent leaves was observed after severe heat waves that occurred when the tassel tissue was about to emerge from the leaf whorl. Thus occurrence of tassel blastinfluences the occurrence leaf firing under heatstress condition due increase in cell injury leads to release of reactive oxygen species complexes and leakage of electrolytes which leads to chlorosis, and death of the tissue under high temperature. Chen *et al.* (2010) also reported that under high temperature stress condition leaf firing reduces photosynthetic apparatus which lead to reduction in grain yield. Maize pollen viability decreases with exposure to temperatures above 35 °C (Dupuis and Dumas, 1990). Das *et al.* (2010) reported that high temperature can delay anthesis and damage most of tassels resulting in little or no pollen production and increasing the occurrence of male sterile plants in field.

SPAD chlorophyll and leaf senescence

In heat stress condition, leaf sensensedue to increase in cell injury leads to release of reactive oxygen species complexes and leakage of electrolytes which leads to chlorosis, and death of the tissue, leading to inappropriate production of assimilates which are required for proper growth. Crafts-Brandner *et al.* (2002) reported reduced content of chlorophyll might cause a drastic reduction in the efficiency of the photosynthetic machinery of crop plants due to reduction in activation of rubisco enzyme alters rate of ribulose-1,5-biphosphate regeneration by disruption of electron transport and inactivation of oxygen evolving enzymes of PS II. Plant senescence is a common physiological phenomenon which leads to drying of the leaves but in heat stress phenomenon proceeds at faster rate moving towards the flag leaf by destroying the chlorophyll content. Guendouz *et al.*(2012) reported that plant having high chlorophyll content showed slow leaf senescence and produces maximum yield under heat stress condition. Renu *et al.*(2004) reported that plant having heat tolerant was characterized by less cell injury percentage under heat stress condition.

Plant and Ear Height

Effect of heat stress was most prominent on plant and ear height reduction might be result of the effect of heat stress on internal –nodal elongation. This research finding was supported by (Weaich *et al.*, 1996; Cairns *et al.*, 2012).

Leaf area index (LAI)

Leaf area index of the decreases was observed after severe heat waves due to leaf growth pattern of maize increases in rang of temperature 0-35ºC with decline at 35-40ºC. Leaf area expansion is of great importance for light interception and for photosynthesis; it varies with the quantity of assimilates allocated to the production of leaves and the ratio of the leaf area produced per unit of leaf dry matter. Heat stress causes translocation of the photosynthetic products cannot fully match the increased rates of carbon fixation under the prevailing conditions, this results in the thickening of the existing leaves and the formation of thicker new leaves, and therefore in a sharp decrease of leaf area in the pre-anthesis period. It can also be noted that the LAI is maximum at tasselling or later and further slight decrease in leaf area is attributed to the senescence of the old (thinner) leaves, so that the younger thicker leaves remain on the stand a and longer growing cycle and normally a higher LAI and determine the overall LAI value. This finding was agreement with earlier finding reported by (Danalatos, N. G., Kosmas, C. S., Driessen, P. M., and Yassoglou, N., 1994). Karim *et al.* (2000) also noted that leaf area and day time leaf expansion rate were good thermo tolerance trait of tropical maize under heat stress condition.

Ear Per Plant (EPP)

Significant increases in the frequency of barrenness in high temperature due reduction in average ears per plant was observed. It is attributed to the fact that different vegetative and reproductive organs undergo active growth at the same stage, which incurs competition for assimilates among organs. Rattalino Edreira *et al.* (2011)finding was agreement with earlier finding reported as changes in distribution of assimilates might be cause for reduced reproductive growth, particularly ears per plant. Cicchino *et al.* (2010b) reported that similar findings has been reported in previous studies in maize crop exposed to high temperature at flowering stage.

Number of kernel per ear (NKE^{-1})

A number of factors could be responsible for reduction in number of kernels ear^{-1} under heat stress condition. Duke & Doehlert, (1996) & Rattalino Edreira *et al.* (2011) finding was agreement with earlier finding reported as number of kernel per ear reduces due to reduced pollen viability and receptivity of silk, increased frequency of kernel abortion, decreased cell division in endosperm, reduced sink capacity of developing kernels, reduced starch grain number and overall starch synthesis, increased soluble sugar accumulation, duration of grain filling, kernel development and enzyme activities. Cicchino *et al.* (2010b) reported similar finding that stress in pre-anthesis stress leading to barrenness in plants, while absorption of fertilized structure and reduced ear growth rate lead to reduction in kernel number and ultimate affect crop yield. Moser *et al.* (2006) reported that stress before and immediately after pollination may lead to failure of number of kernel development ear[-1]. Hussain *et al.* (2010) also reported that, number of rows cob[-1], number of kernels row[-1] and yield per plant were much reduced in spring season due to heat stress.

Silk Receptivity

Pollen shed may occurs for up to 2 weeks but usually lasts for 5 to 8 days with peak shed by about days 3.Silk can grow 2 to 3 inch per day and maximum growth by 3 and 4 days after first silk. Silk longevity is around 10 days and maximum up to 14 days but under heat stress condition desiccate the prematurely will appear as erratic pattern of fertilization along the ear with most fertilized ovule located at the base. A number of factors could be responsible for reduction in number of silk receptivity under heat stress on corn kernel set, seasonal pollen production, silk elongation pattern and duration of silk receptivity. The seasonal pollen production determine kernel per plant at pollen densities less than 3000 pollen grain per silk. It was found that a minimum pollen shed density per exposed silk is required to achieve maximum kernel set and grain yield reported by (Westgate *et al.*, 2003).Silk receptivity can be drastically reduced by as much as 80% during high temperatures due tosudden pollen shedding over a very short time (Fonseca *et al.*, 2005).Anderson *et al.* (2004) found that kernel set and yield stability are impacted by variation among hybrid for silk elongation and senescence. Campos *et al.*, (2004) suggested that selection based on performance in multi-environment trials increased grain yield under drought trough increase yield potential and kernel set, rapid silk exertion and reduced barrenness through at lower rate than under optimal condition.

Shelling percentage

Rowhani *et al.* (2011) reported that significant variation in shelling percentage under heat stress condition might be associated with lower grain yield traits such as pollen viability and fertilization under high temperature. This was because of grain filling period was most sensitive to heat stress as reported by (Thompson, 1986).

Thousand Kernel weight (TKW)

Rise in temperature beyond 30°C impacts the activity of *Rubisco* in maize, which in turn reduces photosynthesis and ultimately decreases grain filling period and grain size (Steven *et al.*, 2002).Kernel weight is influenced by source-sink relationships during grain fill with increased kernel weight being caused by irradiance level, grain-fill duration, and plant and kernel growth rate (Gambin *et al.*, 2006).The reduction of thousand kernel weight in agreement with findings (Abendroth *et al.*, 2011).

Grain Yield

Maize yields have been shown to have an optimum growing temperature of 29 °C and 30 °C, respectively; temperatures above this threshold result in yield decreases (Schlenker & Roberts, 2009).The major effect of high temperature is embryo abortion, which is related to the inhibition of photosynthesis and the subsequent reduction in assimilates available to developing kernels. Exposure to temperatures above 30 °C damaged cell division and amyloplast replication in maize kernels which reduced the size of the grain sink and ultimately yield (Commuri and Jones, 2001).The of the yield decrease up to 100%, larger than those estimated in previous studies (Lobell & Field, 2007; Schlenker & Roberts, 2009). Lobell and Field (2007) showed maize yields decreased 8.3% per 1°C rise without any complicating effect due to water stress. Khodarahmpour *et al.* (2011) observed reduction of grain yield up to 70% under heat stress might be due to low pollen viability, silk receptivity and longer ASI duration in heat stressed condition.

Table 2: Mean square comparison for different traits in 20 inbred lines of maize under high temperature stress at NMRP,Rampur, Chitwan (2016).

Source of variance	df	ASI	LF	TB	LS	SPAD	EH	LAI	EPP
REP.Block	6	0.76ns	29.58ns	10.36ns	0.98ns	8.45ns	24.38ns	0.03ns	0.015ns
Inbred line	19	2.96*	251.3*	369.75**	473.03**	80.83*	264.92**	0.5*	0.16*
Error	13	0.69	92.58	105.7	0.26	29.45	29.34	0.107	0.03

Table 3:Mean square comparison for different traits in 20 inbred lines of maize under high temperature stress at NMRP,Rampur, Chitwan (2016).

Source of variance	df	SR	CD	CL	NKRE	NKE	NKR	SP	TKW	GY
REP.Block	6	2.46ns	0.07ns	0.14ns	0.07ns	0.521ns	0.52ns	2.11ns	91ns	189.9 ns
Inbred line	19	465.98**	4.10**	69.13**	63.3**	86.83**	86.84**	465.98**	39025**	169314**
Error	13	12.96	0.06	0.798	0.23	1.118	1.11	12.96	100	743

Table 4: Mean of 20 maize inbred lines for various traits evaluated under normal (N) and heat stressed condition (S) condition

Inbred	DT N	DT S	DS N	DS S	ASI N	ASI S	LAI N	LAI S	LF% N	LF% S	TB% N	TB% S	LS% N	LS% S
NML 2	71	71.5	74	77	3	5.5	3.5	3.5	5.03	5.6	25.9	58.3	20	50
RL 101	71	71.5	73	75	2	3	3.04	2.61	6.97	14	31.1	36.7	50	60
RL105	73	74.5	75.5	79	2.5	4	2.93	2.84	9.4	39	39.1	46.2	25	65
RL107	68.5	71	70	74	1.5	2.5	2.43	2.42	9.95	11	17.1	27	30	60
RL111	70.5	71.5	73.5	77	3	5.5	2.74	2.29	6.67	24	16.7	48.1	30	50
RL 140	67.5	70	70	73	2.5	3	2.28	2.21	7.88	0	3.3	24	25	75
RML115	70	70.5	73	76	3	5.5	2.65	2.63	7.5	43	21.7	24.9	30	45
RML 17	70.5	71	73	73	2.5	2	2.8	2.7	7.18	14	13.8	17.1	35	80
RML 20	70.5	71.5	73	75	2.5	3	3	2.85	7.5	14	17.1	17.4	30	85
RML 24	71.5	72	70.5	77	-1	5	3.2	3.09	6.67	28	23.3	55.8	35	75
RML 32	68.5	71.5	71	75	2.5	3	2.44	2.33	8.33	18	18	22.7	30	60
RML4	71	71	74	77	2.5	5.5	2.66	2.65	7.74	28	30.4	52.3	25	50
RML 40	67	68	68	71	1	3	2.83	2.8	7.18	10	3.8	43.3	25	80
RML 57	69.5	71.5	71.5	74	2	2.5	2.5	2.33	8.12	10	9.1	37.6	25	60
RML 7	67	67	68.5	70	1.5	3	2.31	2.29	12.7	14	31.5	35.7	60	100
RML 76	70.5	70	72	75	1.5	4.5	2.09	1.75	7.18	13	0	23.3	40	45
RML 86	72	71.5	75.5	77	3.5	5.5	3.22	3.07	4.17	21	18.3	50	30	65
RML 91	68.5	70.5	71	74	2.5	3	3.17	3.15	6.9	0	3.6	16.7	35	40
RML 95	68.5	71	72	75	2.5	4	2.7	2.62	7.88	22	28.2	44.5	25	70
RML 96	70	72	72.5	75	2.5	3	2.91	2.86	6.67	14	6.7	31	35	60
G.Mean	69.9	71	72.1	75	2.2	3.8	2.77	2.64	7.58	18	18	35.6	32	63.8
F –test	NS	NS	*	*	*	*	*	*	NS	*	*	**	NS	**
CV%	2.25	2.85	2.17	2.45	37	21.9	14	12	38.01	52.3	40.3	22.93	32	12.4
LSD	3.4	4.37	3.39	3.97	1.7	1.8	0.88	0.70	6.22	20.8	15.68	17.64	22.1	17.2

Table 5 : Mean of 20 maize inbred lines for various traits evaluated under normal (N) and heat stressed condition (S) condition

Inbred	SPAD		PH (cm)		EH (cm)		PM (Days)		EPP		CD (cm)		CL(cm)	
	N	S	N	S	N	S	N	S	N	S	N	S	N	S
NML 2	46.65	41.59	176.3	147.8	96	81	115.5	106.5	1.436	1.042	3.41	0	10.92	0
RL 101	43.975	44.53	152.5	133.3	85.5	73.8	108	103	1.5	1.25	3.19	2.67	9.70	8.24
RL105	41.1	24.5	146.7	141.8	86.5	80	111.5	108	1.267	0.885	2.75	0	11.4	0
RL107	50.125	43.38	143.9	129.2	68	65.8	114.5	105	1.7	1.339	3.99	2.99	12.95	12.43
RL111	36.93	24.27	164.5	148.3	86.1	80.6	120	104.5	1.95	2.176	3.87	0	12.19	0
RL 140	41.215	37.45	136.8	127	72.2	60.2	121	107.5	1.406	1.14	3.9	2.67	13.30	12.4
RML115	38.42	35.72	108.3	102.6	61.6	54.5	121.5	109	1.042	0.862	3.69	0	7.27	0
RML 17	47.375	44.11	159.2	136.6	86.7	74.5	114	105.5	1.419	0.893	3.83	2.43	12.40	11.62
RML 20	44.26	39.87	133.8	118	68.2	52.8	114	103	1.550	1.488	3.56	2.89	14.10	12.84
RML 24	45.65	27.31	153.6	144.3	93.3	77.6	119.5	108	1.333	1.196	3.49	0	9.96	0
RML 32	48.14	43.18	120	113.4	46.5	43.5	114	108	1.250	1.045	3	2.83	13.05	11.96
RML4	44.7	40.25	125.6	116.7	72.4	54	118	109	1.435	1	3.1	0	10.75	0
RML 40	46.725	41.25	134.9	125.8	63.7	60.2	116.5	106	1.254	1.033	4.06	3.65	12.10	11.36
RML 57	48.24	45.56	134.1	128.2	66.5	60.9	116	104.5	1.350	1.143	3.29	2.45	12.95	0
RML 7	38.78	36.41	148.6	135.8	74.7	67.3	115	100.5	1.30	1	3.87	3.02	12.18	12.08
RML 76	44.88	41.1	123.9	115	56.8	51.4	115	108.5	1.3	1.1	3.44	2.32	11.13	9.53
RML 86	37.605	33.48	126.8	120.5	74.8	69.4	114	107	1.35	1	3.53	0	13.02	0
RML 91	40.33	42.45	159.8	150	95.5	83.4	120	109	1.55	1.3	3.16	3.08	13.85	13.27
RML 95	37.4	38.55	142.9	128	77	71.5	116	108.5	1.7	1.324	3.57	0	10.55	0
RML 96	41.53	38.32	155.4	134.8	75.8	69.6	119.5	108.5	1.533	1.143	3.73	2.54	12.60	11.83
GM	43.202	38.16	142.4	129.9	75.4	66.6	116.17	106.5	1.431	1.168	3.522	1.677	11.82	6.89
F –test	**	*	**	*	**	**	**	**	NS	*	*	**	*	**
CV%	1.17	14.22	5.2	7.91	9.25	8.13	1.45	1.5	14.2	15	7.5	14.66	8.16	12.97
LSD	1.0941	11.723	16.02	22.21	15.08	11.7	3.654	3.432	0.441	0.379	0.5708	0.532	2.086	1.931

Table 6 : Mean of 20 maize inbred lines for various traits evaluated under normal (N) and heat stressed condition (S) condition

Inbred	NKRE		NKE		NKR		SR %		SP %		TKW (g)		GY (kg/ha)	
	N	S	N	S	N	S	N	S	N	S	N	S	N	S
NML 2	10.6	0	74.4	0	10.6	0	54.74	0	45.8	0	259	0	1457	0
RL 101	13.5	12.6	87	53	14.7	12.5	46.73	30.55	60.5	57.95	289	264.6	1469	551.9
RL105	13	0	92.5	0	13	0	47.01	0	17.4	0	257	0	366	0
RL107	12.30	10.6	165	50.2	15	12.8	70.44	22.35	51.9	18.48	288	270.6	1918	264
RL111	12.2	0	173.9	0	12.2	0	77.7	0	48.8	0	261	0	2153	0
RL 140	12.8	11.7	210.4	110.5	17	14.8	79.27	36.77	61.4	54.65	289.5	279.6	2352	702.9
RML115	12.2	0	92.4	0	12.2	0	53.35	0	30.2	0	269	0	818	0
RML 17	12.2	10	187.8	67.9	14.6	12.4	78.1	31.9	71.3	58.1	275	271.1	2201	565.3
RML 20	11.6	10.4	115.5	50.3	15.4	13.2	55.37	20	40.4	21.62	316	306.1	1213	273.6
RML 24	12.4	0	104.9	0	12.4	0	54.73	0	38.3	0	229	0	1196	0
RML 32	10.5	10	74.3	45.2	14.4	12.2	53.41	26.3	54.9	42.3	306	293	1130	536.7
RML4	11.2	0	92.5	0	11.2	0	62.06	0	39.3	0	269.5	0	1287	0
RML 40	11.	10.1	130.5	74.8	13.6	12.4	74.52	34.27	62.3	56.5	301	276	2018	643
RML 57	11.8	11.3	143.7	48.7	15.5	13.8	70.57	22.95	51.1	36.22	253	204.7	1903	342.6
RML 7	12.6	12.6	102.5	70	14.2	12	50.84	26.7	55.9	51.2	333	320	582	526.5
RML 76	14.2	10.8	124.6	94	16.5	14.3	68.19	34.57	79.0	52.12	272	255.7	1689	689.5
RML 86	12.4	0	108.2	0	12.8	0	61.75	0	44.3	0	257	0	1577	0
RML 91	12.8	12.8	275	130.2	16	13.8	79.39	38.2	60.4	40.3	289	276	2346	716.8
RML 95	11.2	0	143.4	0	11.2	0	76.8	0	54.1	0	169.4	0	1991	0
RML 96	12	10	118	46.2	14.4	12.2	77.25	20.06	56	46.3	289	277.9	2090	508
GM	12.24	6.64	130.8	42.1	13.8	7.82	64.61	17.23	51.2	26.79	273.3	164.8	1588	316
F –test	Ns	**	**	**	*	**	**	**	**	**	**	**	**	**
CV%	9.5	7.2	14.31	21	9.58	13.52	4.04	11.57	13.40	14.7	6.702	6	13.28	8.62
LSD	2.166	1.037	40.45	19.21	2.86	2.284	5.652	4.31	14.83	8.56	39.73	21.61	455.8	58.89

Coefficient of Phenotypic and Genotypic Variation (PCV and GCV)

Phenotypic coefficient of variation (PCV) and genotypic coefficient of variation (GCV) useful for comparing the relative amount of phenotypic and genotypic variations among different traits and very useful to estimate the scope for improvement by selection (Table7).PCVs were slightly higher than GCVs for all selection parameters indicating presence of environmental influence on the expression of character. High value of PCV and GCV observed in anthesis silking interval, leaf firing, ear per plant, silk receptivity, cob diameter, cob length, number of kernel ear^{-1}, number of kernel ear^{-1},shelling percentage, thousand kernel weight and grain yield under heat stress condition whereas for antheis silking interval, tassel blast, number of kernel ear^{-1}, shelling percentage and grain yield showed that selection can be effective for these traits but also indicated the existence of substantial variability, ensuring ample scope for their improvement through selection. This observation was confirmed with finding reported by Rafique *et al.* (2004) and Rafiq *et al.*(2010).On the other hand, low both coefficient recorded for days to 50% tasseling, silking, plant physiological maturity in   heat stress   whereas days to 50% tasseling ,silking, plant physiological maturity, number of kernel ear^{-1} in normal condition indicated existence of low variability among the genotypes was very low for these traits under given condition.

Heritability (H²b) and Genetic advance

The considerable difference in heritability value for different character was observed in normal and heat stress condition. According to Singh, (2001), heritability of a trait is considered as very high when the value is 80% or more and moderates when it ranged from 40-80% and when it is less than 40%, it is low. High magnitude of board sense heritability estimated in all character thousand kernel weight, grain yield, shelling percentage, number of kernel ear^{-1},ear height in heat stress whereas SPAD reading, grain yield, number of kernel ear^{-1},shelling percentage, plant height and thousand kernel weight in normal condition indicated possibility of effective selection for genetic improvement of these traits. On the other hand, moderate heritability estimates for cob diameter, leaf firing, plant height, pad reading, days to 50 % silking in heat stress condition as compared to day to 50 % silking and tasseling, anthesis silking interval, cob diameter, number of kernel ear^{-1}, leaf area index in normal condition which suggested that selection should be delayed to more advance generations for these traits. But ear height and leaf senescence were genotype specific response they showed non-significant response with grain yield. Similar finding for leaf senescence was reported by (Hussain *et al.*, 2006). Due to high heritability and genetic advance 5 % selection intensity for traits grain yield, number of kernel ear^{-1}, silk receptivity, shelling percentage, and thousand kernel weight indicated presence of additive gene effect and early selection and could be used as target traits to improve grain yield under both condition. Anthesis silking interval, leaf firing, tassel blast, leaf senescence, ear per plant, leaf area index, cob length, number of kernel row ear^{-1} showed   low genetic advance and moderate heritability means these traits were govern by non-additive or dominant gene action and suggested as hybrid plants have higher capacity to tolerate heat stress in field conditions than their parents. These result find supported from earlier studies by Bello *et al.,*(2012) that there was greater magnitude of broad sense heritability and high genetic advance in grain yield, number of kernel ear[-1] and ear height. Sumathi, P., Nirmalakumari, A., and Mohanraj, K. (2005) reported that selection of this parameter could be used for improvement of grain yield. Betran *et al.* (2003) reported similar result for genetic improvement of maize under stress condition can be achieved by selection of grain yield and traits tassel blast, leaf firing leaf senescence, anthesis silking interval, crop maturity period ear per plant, plant height.

Table 7 : Genetic parameter of maize inbred lines under heat stress and normal condition

Traits	Heritability		GA at 5 %		Correlation	
	Normal	Stress	Normal	Stress	Normal	Stress
GY	0.88	0.98	1080.84	475.4	-	-
DS (day)	0.55	0.46	2.62	2.34	-0.221ns	-0.766**
DT(day)	0.36	0.06	1.48	0.24	-0.343ns	-0.492*
ASI (day)	0.49	0.62	1.13	1.73	0.096ns	-0.726**
LAI(cm2)	0.45	0.65	0.5	0.74	0.06ns	-0.445*
SPAD	0.98	0.47	8.34	7.13	0.085ns	0.560*
EH (cm)	0.76	0.8	22.02	20	0.124ns	0.249ns
PH (cm)	0.82	0.52	30.1	16.02	0.293ns	0.041ns
LF	0	0.6	0	2.07	-0.123ns	-0.692*
TB	0.62	0.72	2.5	1.86	-0.764**	-0.679*
LS	0.2	0.75	0.41	1.58	-0.24ns	0.167ns
PM (day)	0.78	0.64	5.8	3.53	0.327ns	-0.162ns
EPP	0.23	0.69	0.1	0.45	0.463*	-0.132ns
SR %	0.95	0.88	23.41	16.4	0.911**	0.98**
CD (cm)	0.57	0.44	0.47	1.53	0.439ns	0.896**
CL (cm)	0.62	0.62	2.25	6.25	0.51*	0.857**
NKRE	0.32	0.78	0.83	7.29	-0.2ns	0.917**
NKE	0.87	0.84	94.17	40.43	0.74**	0.941**
SP %	0.84	0.94	19.3	25.58	0.941**	0.963**
TKW (g)	0.75	0.99	56.25	286.65	-0.362ns	0.94**

Correlation analysis

The grain yield had positive and significant phenotypic correlation with silk receptivity, shelling percentage, cob length and diameter, number of kernel ear[-1], number of kernel row ear[-1], number of kernel ear[-1],SPAD chlorophyll, and thousand kernel weight whereas it was significant and negative correlation with tassel blast, anthesis silking interval, leaf area index, leaf firing under heat stress condition as shown in Table 8. Khodarahmpour & Choukan (2011) reported similar finding in a study of fifteen inbred line under heat stress condition reported that grain yield had positive and significant correlation with number of kernel ear[-1], no of kernel row, no of kernel ear[-1], 1000-grain weight, and cob diameter. Kaur *et al.* (2010) reported similar result significant negative association leaf firing and tassel blast with grain yield under heat stress. Cairns *et al.*(2012) reported significant correlation between grain yield and ASI and SPAD reading was negatively correlated. Krasensky, J., and Jonak, C. (2012) reported similar result for significant positive for association between chlorophyll content and grain yield under drought stress. Betran *et al.*(2003a) reported similar result for SPAD chlorophyll content and EPP with grain yield under drought stress but relation of EPP was found non-significant.

Table 8: Pearson's Correlation coefficient among different traits under heat stress condition at NMRP, Rampur (2016).

	ASI	LF	TB	LS	SPAD	EH	LAI	EPP	SR	CD	CL	NKRE	NKE	NKR	SP	TKW
ASI	1															
LF	.546[*]	1														
TB	.641[**]	.328	1													
LS	-.442	-.113	-.031	1												
SPAD	-.555[*]	-.607[**]	-.474[*]	-.063	1											
EH	.109	-.054	.389	-.051	-.386	1										
LAI	.345	.101	.474[*]	.021	-.238	.508[*]	1									
EPP	.094	-.137	.052	-.140	-.282	.251	-.161	1								
SR	-.788[**]	-.723[**]	-.714[**]	.195	.617[**]	-.246	-.457[*]	-.093	1							
CD	-.860[**]	-.687[**]	-.686[**]	.317	.629[**]	-.311	-.411	-.039	.937[**]	1						
CL	-.758[**]	-.626[**]	-.761[**]	.320	.496[*]	-.260	-.339	-.026	.869[**]	.907[**]	1					
NKRE	-.852[**]	-.705[**]	-.715[**]	.258	.640[**]	-.266	-.464[*]	-.038	.954[**]	.972[**]	.871[**]	1				
NKE	-.682[**]	-.748[**]	-.714[**]	.117	.513[*]	-.153	-.396	-.052	.953[**]	.853[**]	.831[**]	.884[**]	1			
NKR	-.848[**]	-.710[**]	-.745[**]	.234	.648[**]	-.333	-.488[*]	-.038	.956[**]	.968[**]	.873[**]	.987[**]	.887[**]	1		
SP	-.763[**]	-.630[**]	-.605[**]	.292	.572[**]	-.236	-.463[*]	-.180	.945[**]	.883[**]	.803[**]	.910[**]	.846[**]	.901[**]	1	
TKW	-.854[**]	-.668[**]	-.743[**]	.340	.609[**]	-.330	-.446[*]	-.047	.932[**]	.982[**]	.937[**]	.978[**]	.850[**]	.975[**]	.898[**]	1
GY	-.726[**]	-.692[**]	-.679[**]	.167	.560[*]	-.237	-.445[*]	-.132	.980[**]	.896[**]	.857[**]	.917[**]	.941[**]	.917[**]	.963[**]	.904[**]

Values are significant difference at 5 % level of significance (*) and highly significant at 1 % level of significant (**), ASI= AnthesisSilkinginterval,LF %= leaf firing%,TB= Tassel blast,LS=Leaf senescence,SPAD=SPAD chlorophyll,EH=Ear height,PH=Plant height,LAI= leaf area index,EPP= Ear per plant,SR%= Silk receptivity,CD=Cob diameter,CL=Cob length,NKRE=Number of kernel per row,NKE=Number of kernel per Ear,NKR=Number of kernel per row,SP%= Shelling percentage, TKW=Thousand kernel weight(g),GY=GrainYield(kg ha^{-1}).

Path coefficient analysis

Direct effect of thousand kernel weight (0.786), shelling percentage (0.552), number of kernel ear^{-1} (0.448), silk receptivity (0.279), leaf area index (0.058) on grain yield had highest positive value as compared to all other traits such as ear per plant (0.011) exerted positive direct effect on yield based on direct effect in path analysis but it is non-significant correlated with grain yield hence may not be statistically considerable. El-Badawy *et al.*(2011) reported similar result highest positive direct effect of 100 grain weight on yield per plant in maize. Krishnaji *et al.* (2017) found similar result high positive direct effect of shelling percentage and number of kernel per ear under heat stress condition. Jawaria Azhar, J., J.,Ramzan & Ahmad, R. M. (2016) reported similar result high positive direct effect of thousand kernel weight on grain yield under heat stress condition in maize. Similarly correlation coefficient of SPAD chlorophyll, cob diameter, cob length, Number of kernel row ear^{-1},number of kernel row were positive and significant with grain yield while there direct effects on grain yield were negative. But negative direct effects of these traits were nullified by their positive indirect contribution via other yield components. Similarly correlation coefficient of leaf area index was negative and significant with grain yield while there direct effects on grain yield were positive. But positive direct effects of these traits were nullified by their negative indirect contribution via other yield components. On contrary, some character of anthesis silking interval and tassel blast exerted positive direct effect on grain yield. However positive direct effect of these traits was nullified by their negative indirect contribution via other yield components. Thousand kernel weight showed the highest positive indirect contribution towards grain yield via shelling percentage (0.4960), number of kernel per ear (0.381), silk receptivity (0.260), anthesis silking interval (0.126), ear height (0.0380) and leaf firing(0.015).However, it showed negative indirect effect via tassel blast, leaf senescence, SPAD chlorophyll, leaf area index, ear per plant, cob diameter and length, number of kernel row per ear and number of kernel per row respectively. Shelling percentage exhibited had positive indirect contribution on grain yield via thousand kernel weight, number of kernel per ear, silk receptivity, ear height, leaf firing and anthesis silking interval. However, it showed negative indirect effect via tassel blast, leaf senescence, SPAD chlorophyll, leaf area index, ear per plant, cob diameter & length, number of kernel row per ear and number of kernel per row respectively. Number of kernel per ear exhibited had positive indirect contribution on grain yield via thousand kernel weight, shelling percentage, silk receptivity, anthesis-silking interval, ear height, and leaf firing. However, it showed negative indirect effect via tassel blast, leaf senescence, SPAD chlorophyll, leaf area index, ear per plant, cob diameter and length, number of kernel row ear^{-1} and number of kernel row^{-1}respectively. Silk receptivity exhibited had positive indirect contribution on grain yield via thousand kernel weight, shelling percentage, anthesis silking interval, ear height, and leaf firing. However, it showed negative indirect effect via tassel blast, leaf senescence, SPAD chlorophyll, leaf area index, ear per plant, cob diameter and length, number of kernel row ear^{-1} and number of kernel row^{-1} respectively.The negative correlation between grain yield and anthesis silking interval, tassel blast and leaf area index indirect influence via thousand kernel weight and shelling percentage, silk receptivity and number of kernel ear^{-1}.This traits needs consideration, because direct effect these traits positive in direction. The positive correlation between grain yield and SPAD chlorophyll, cob diameter, cob length, number of kernel row[-1] and number of kernel row indirect influence via thousand kernel weight and shelling percentage, silk receptivity and number of kernel ear[-1].This trait needs consideration, because direct effect these traits negative in direction(Table 9).

Table 9: Direct (diagonal) and indirect effects of different traits on grain yield.

	ASI	LF	TB	LS	SPAD	EH	LAI	EPP	SR	CD	CL	NKRE	NKE	NKR	SP	TKW
viaASI	**0.022**	-0.081	-0.095	0.065	0.082	-0.016	-0.051	-0.014	0.116	0.127	0.112	0.126	0.101	0.125	0.113	0.126
via LF	-0.012	**-0.022**	-0.007	0.003	0.014	0.001	-0.002	0.003	0.016	0.015	0.014	0.016	0.017	0.016	0.014	0.015
viaTB	0.062	0.032	**0.096**	-0.003	-0.046	0.037	0.046	0.005	-0.069	-0.066	-0.073	-0.069	-0.069	-0.072	-0.058	-0.072
viaLS	0.077	0.020	0.005	**-0.173**	0.011	0.009	-0.004	0.024	-0.034	-0.055	-0.055	-0.045	-0.020	-0.041	-0.050	-0.059
viaSPAD	0.047	0.052	0.040	0.005	**-0.085**	0.033	0.020	0.024	-0.053	-0.054	-0.042	-0.054	-0.044	-0.055	-0.049	-0.052
viaEH	-0.013	0.006	-0.045	0.006	0.044	**-0.115**	-0.058	-0.029	0.028	0.036	0.030	0.031	0.018	0.038	0.027	0.038
viaLAI	0.020	0.006	0.027	0.001	-0.014	0.029	**0.058**	-0.009	-0.026	-0.024	-0.020	-0.027	-0.023	-0.028	-0.027	-0.026
viaEPP	0.001	-0.001	0.001	-0.002	-0.003	0.003	-0.002	**0.011**	-0.001	0.000	0.000	0.000	-0.001	0.000	-0.002	-0.001
viaSR	-0.220	-0.202	-0.199	0.054	0.172	-0.069	-0.128	-0.026	**0.279**	0.262	0.243	0.267	0.266	0.267	0.264	0.260
viaCD	0.202	0.161	0.161	-0.074	-0.148	0.073	0.097	0.009	-0.220	**-0.235**	-0.213	-0.228	-0.200	-0.227	-0.207	-0.231
viaCL	0.056	0.046	0.056	-0.024	-0.036	0.019	0.025	0.002	-0.064	-0.067	**-0.073**	-0.064	-0.061	-0.064	-0.059	-0.069
viaNKRE	0.243	0.201	0.204	-0.074	-0.182	0.076	0.132	0.011	-0.272	-0.277	-0.248	**-0.285**	-0.252	-0.281	-0.259	-0.279
viaNKE	-0.305	-0.335	-0.320	0.052	0.230	-0.069	-0.177	-0.023	0.427	0.382	0.372	0.396	**0.448**	0.397	0.379	0.381
viaNKR	0.358	0.300	0.314	-0.099	-0.273	0.140	0.206	0.016	0.403	0.408	-0.368	-0.416	-0.374	**-0.422**	-0.380	-0.411
via SP	-0.422	-0.348	-0.334	0.161	0.316	-0.130	-0.256	-0.099	0.522	0.488	0.444	0.503	0.467	0.498	**0.552**	0.496
via TKW	-0.671	-0.525	-0.584	0.267	0.479	-0.259	-0.351	-0.037	0.733	0.772	0.737	0.769	0.668	0.766	0.706	**0.786**
SUM	-.726**	-.692**	-.679**	0.1674	.560*	-0.237	-.445*	-0.132	.980*	.896*	.857**	.917**	.941**	.917**	.963**	.904**

Stress tolerance indices

The analysis of variance showed significant difference between inbred lines (Table 10). Among the inbred line RL-140 and RML-91 produced the high grain yield under both stress and normal condition.RML-17 and RML-111 relatively high yield in normal condition, but relatively low yield under stress condition. RL-105 and RML-115 had lower yielded under normal condition and zero yield under stress condition. In contract, RML-76 and RML-40 had a relatively high yield under heat stress condition but relatively intermediate yield under normal condition. Eight genotype NML-2, RL-105, RL-111, RML-115, RML-24, RML-4, RML-86, RML-95 lines had produces barren cob under stress condition (Table 10).

Table10 : ANOVA of stress tolerance indices and yield in heat stress and normal condition in maize inbred line at NMRPRampur, Chitwan (2016).

Source of variance	df	YP	YS	SSI	STI	TOL	GMP	MP
				Mean of squares				
REP.Block	6	28379ns	2462ns	0.00092ns	0.00590ns	24975ns	4467ns	6389ns
Inbred line	19	673050*	169506*	0.167**	0.107**	572894*	519707*	278092*
Error	13	44506	8247	0.005	0.0032	47914	13638	17362

*** *,ns,Significant at 1% levels, respectively, ns= Non-significant.Yp=Yield in non-stress conditions, Ys=Yield in stress conditions, GMP=Geometric Mean Productivity, MP= Mean Productivity, SSI=Stress Susceptibility Index, STI=Stress Tolerance Index, TOL= Tolerance index.*

SSI is the ration between the traits at optimum and those at heat stress condition a higher SSI value indicate lowest trait value stressed condition and vice versa. According to Khanna-Chorpa and Vishwanathan (1999), the genotype having SSI value less than 0.5 are highly tolerant, those having values between 0.5 to less than 1 are moderately tolerant and those having values equal or more than 1 are susceptible heat stress. Based on the STI, lines RML-17 (0.11),RML-32(0.64) and RML-76(0.73) inbred lines showed lowest SSI revealed as highly tolerance to heat stress condition whereas RL-101 (0.7) , RL-140 (0.8) , RML-17 (0.92) , RML-20 (0.95) , RML-32 (0.64) , RML-40 (0.84) , RML-76 (0.40) , RML-91 (0.8) , RML-96 (0.9) lines having values between 0.5 to less than 1 and revealed as moderately tolerance. Similarly NML-2, RL-105, RL-111, RML-115, RML-24, RML-4, RML-86, RML-95 all had 1.25, RML-57 (1) and RL-107 (1) those are highly susceptible in nature.

The higher value of stress tolerance index was exhibited by RML-17 (0.90) fallowed by RML-32 (0.48) and RML-76(0.40) whereas zero value of STI was found for NML-2,RL-105,RL-111,RML-115,RML-24,RML-4,RML-86,RML-95.High STI revealed as highest tolerance to heat stress while those with least value are considered to be comparatively susceptible.RML-76 produced the moderately yield in both condition.

Based on the TOL index allowed us to select RL-111 (2153) fallowed by RML-95 (1991) and RL-107(1654) inbred as tolerance genotype. RL-111 and RML-95 lines showed zero yield under heat stress. This due to low difference between the two conditions, which decrease the value of the TOL index. Therefore, low TOL does not mean high yielding and genotype yielding should be taken in consideration in addition to this criterion.

Mean productivity index, RML-91(1531) fallowed by RL-140 (1527) were revealed as tolerance Table 3. Therefore, according to these results selection based on MP will improve mean yield under both condition but does not allow discriminating line having high yield under both stress and line having high yield under normal and lower yield under stress condition.
The study of geometric mean productivity showed more comprehensive result. Based on this index, RML-91 (1294.8) fallowed by RL-140(1285.1) were revealed as tolerance and had high yield under both condition as shown in Table 11.

Table 11: Yield under non-stress, yield under stress, Stress susceptible index, Stress Tolerance index, Tolerance index, Geometric mean productive index and Mean Productive index of maize inbred used at NMRP, Rampur, Chitwan (2016).

Inbred/ Indices	Yp	Ys	SSI	STI	TOL	GMP	MP
NML 2	1457	0	1.25	0	1457	0	728
RL 101	1469	551.9	0.7713	0.3829	917	898.5	1011
RL 105	366	0	1.25	0	366	0	183
RL 107	1918	264	1.078	0.1373	1654	706.8	1091
RL111	2153	0	1.25	0	2153	0	1077
RL140	2352	702.9	0.876	0.2996	1649	1285.1	1527
RML 115	818	0	1.25	0	818	0	409
RML 17	2201	565.3	0.9269	0.2585	1636	1114.5	1383
RML 20	1213	273.6	0.9587	0.233	939	572.7	743
RML24	1196	0	1.25	0	1196	0	598
RML 32	1130	536.7	0.6474	0.4821	593	777.6	833
RML 4	1287	0	1.25	0	1287	0	644
RML 40	2018	643	0.8488	0.321	1375	1138.3	1331
RML 57	1903	342.6	1.0251	0.1799	1561	807.4	1123
RML 17	582	526.5	0.1189	0.9049	56	553.6	554
RML 76	1689	689.5	0.7376	0.4099	999	1078.7	1189
RML 86	1577	0	1.25	0	1577	0	788
RML 91	2346	716.8	0.8646	0.3083	1629	1294.8	1531
RML 95	1991	0	1.25	0	1991	0	996
RML 96	2090	508	0.9454	0.2437	1581	1029.9	1299
Grand Total	1588	316	0.9899	0.2081	1272	562.9	952

The result showed that based on MP and GMP inbred line RL-140 and RML-91 were most tolerance while STI and SSI showed RML-17,RML-32 were most tolerance so that need to determine the most desirable stress tolerance criteria, the correlation coefficient between Yp,Ys and quantitative indices of stress tolerance were calculated (Table 12).There are significant correlation between Yp and (MP, TOL, GMP) and between Ys and SSI, STI, GMP and MP and indices GMP and MP consequently appeared as better predictors of Yp and Ys than TOL, SSI and STL. The relationship between both Yp and Ys and MP are consistent with those reported by Fernandez (1992) in Mungbean and Farshadfar and Sutka (2002) in maize. In present study correlation coefficient between SSI and Ys were r= -0.768. The computed r value -0.768 indicate that $(-0.768)^2 \times 100 = 58.98$ % of variation in the mean yield in stress condition is accounted by SSI index. Thus selection for SSI should give decreased yield under heat stress condition. Likewise, correlation coefficient between STI and Ys r= 0.768 indicate it should give a positive yield response in a hot environment. Therefore, high correlation coefficient between Ys and STI and negative correlation coefficient between Ys and SSI indicated that selection for tolerance based on STI and SSI would be worthwhile only when target environment is heat stressed.

The correlation between GMP and MP yield in stress and normal condition were highly positive response and significant especially under stressed condition. Hence, selection for high GMP and MP should give positive response in both environments. According to the result, the use of STI, GMP and MP indices should help to improve heat tolerance in inbred line. MP that showed high positive correlation with grain yield in both stress and non-stress environment should be more efficient in inbred line selection. In general, selection of inbred line maize based on GMP might allow us to improve heat tolerance and potential yield under both environments. Based on GMP indices the RL-140, RML-91 appeared as having high yield potential and low stress susceptibility. Based on result of this study inbred RL-140, RML-91 should be use a source of heat tolerance for breeding program.

Table 12: Phenotypic correlation coefficients between maize inbred lines yield in stress and non-stress conditions and heat stress tolerance at NMRP, Rampur,Chitwan (2016).

YpYs	Ys	SSI	STI	TOL	GMP	MP
Ys	0.399^{ns}					
SSI	0.093^{ns}	-0.768^{**}				
STI	-0.093^{ns}	0.768^{**}	-1.0^{**}			
TOL	0.867^{**}	-0.111^{ns}	0.519^{*}	-0.519^{*}		
GMP	0.536^{*}	0.966^{**}	-0.616^{*}	0.616^{*}	0.056^{ns}	
MP	0.934^{**}	0.701^{**}	-0.227^{ns}	0.227^{ns}	0.631^{*}	0.794^{**}

*and **, Significant at 5% and 1% levels, respectively, ns= Non-significant. YpYield in non-stress conditions, YsYield in stress conditions, GMPGeometric Mean Productivity, MP Mean Productivity, SSIStress Susceptibility Index, STIStress Tolerance Index, TOL Tolerance Index.*

Table 13: Principal component of first and second tolerance indices and yield in stress and non-stress condition in maize inbred at NMRP, Rampur,Chitwan (2016).

Inbred	PC	Eigen Value	Proportion	Cum Proportion	YpYs	SSI	STI		GMP	MP
Line	PC1	3.9144	0.559	55.9	-0.26	-0.49	0.38	-0.38	-0.48	-0.39
	PC2	2.8071	0.401	96.0	0.5	-0.69	0.36	-0.36	**0.043**	**0.36**

From Principal component analysis showed that first PC explained 55.9 % variation and second PC explained 40.1 % (Table 13). Considering the high and negative value of indices of this PCA and high and positive value of indices in this second PC first PC named as low yielding and stress susceptible and second PC as yield potential and heat tolerance. Thus selection of genotype having low PC1 and high PC2 are suitable for both stress and normal condition. Thus inbred RML-91, RL-140, RML-40, RML-17, RML-96, RML-7, RL-107 exhibited low PC1 and high PC2 are superior for both environment conditions. RML-15, RML-115 and RL-105 having high PC1 value and low PC2 value showed poor yield under both environmental condition. Thus result obtained from biplot graph confirmed the correlation analysis (Table 13).

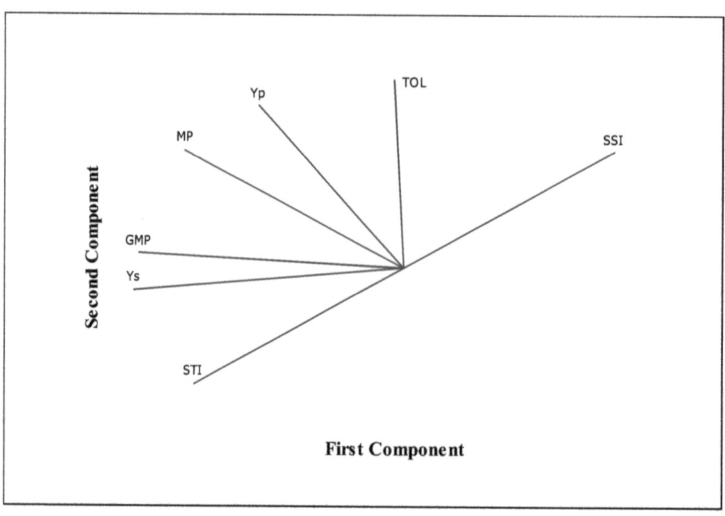

Fig 2: The Biplot display of heat tolerance indices

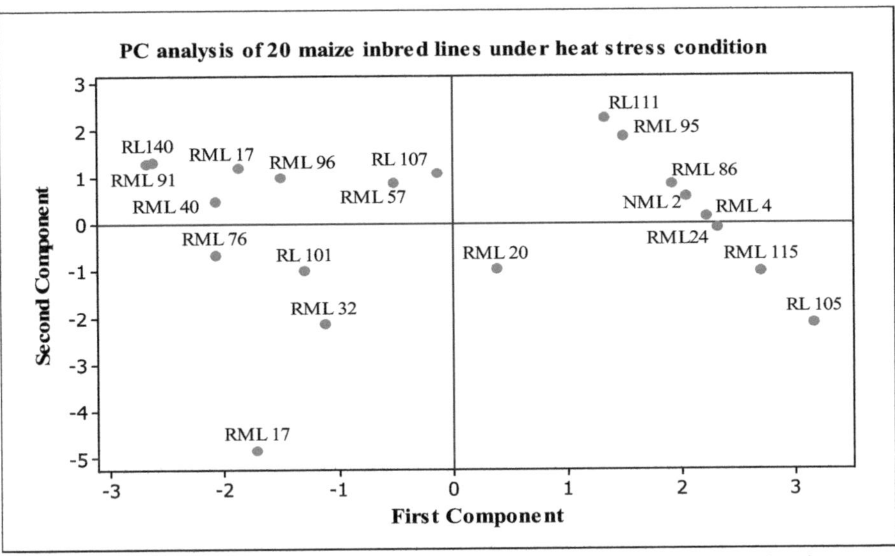

Fig 3: The Biplot display of yield in seven heat tolerance based on the first and second main component of maize inbred lines at NMRP, Rampur,Chitwan (2016).
Cluster analysis

The critical examination of dendrogram reveled four clusters with minimum of 22.47 % similarity level in UPGMA Clustering. The cluster was divided into two groups: group A and Group B consisted of three clusters namely Cluster2, Cluster 3 and cluster 4 and group A contained only one cluster Cluster 1.Cluster 1 consisted of 8 genotypes (NML-2, RL-105, RML-24, RL-111, RML-4, RML-86, RML-95 and RML-115) which represent 40% of total genotypes. Inbred lines grouped in this cluster had longer anthesis silking interval, with maximum tassel blast and leaf firing along with zero value for grain yield including cob length, cob diameter and length, number of kernel row ear^{-1}, number of kernel ear^{-1},number of kernel row^{-1},shelling percentage, silk receptivity and thousand kernel weight. Cluster 2 consisted of 5 genotypes (RL-101, RML-17, RML-32, RML-96 and RML-7), 25 % of total genotypes was characterized with had highest leaf senescence fallowed by thousand kernel weight and lowest for cob diameter and length, ear per plant, number of kernel row ear^{-1},number of kernel ear^{-1} and number of kernel row^{-1}.The inbred lines categorized into cluster 3 (RL-107, RML-20 &RML-57), 15 % of total genotypes had shorter plant and ear height, late physiological maturity and highest for ear per plant. Cluster 4 consisted of 4 (RL-140, RML-76, RML-91 and RML-40), 20 % of total genotypes were characterized by lowest value of tassel blast, leaf firing, leaf area index with highest value of cob diameter and length, ear per plant, number of kernel row ear^{-1}, number of kernel ear^{-1},number of kernel row^{-1}, shelling percentage, silk receptivity and grain yield in heat stress condition. Since this cluster of genotype had superior trait value for heat stressed condition, this genotype may be of interest to researcher. The distance between the clusters centroid was found highest between clusters 1 and 4 and lowest between clusters 2 and 4.

Table 14: Mean of Clustering of 20 Maize inbred lines under heat stress condition at NMRP, Rampur, Chitwan (2016).

Variable	Cluster 1	Cluster 2	Cluster 3	Cluster 4	Centroid
No. of inbred lines	8	5	3	4	20
Anthesis silking interval	5.06	2.8	2.66	3.37	3.8
Leaf area index	2.83	2.55	2.53	2.47	2.64
SPAD Chlorophyll	33.209	41.31	42.93	40.56	38.16
Leaf firing	26.35	14.88	11.6	**5.75**	17.13
Tassel blast	47.51	28.64	27.33	**26.825**	35.63
Plant height	131.25	130.78	125.133	129.45	129.85
Ear height	71.06	65.74	59.833	63.8	66.59
Physiological maturity	107.563	105	104.16	**107.75**	106.47
Leaf sensense	58.75	72	68.33	60	63.75
Ear per plant	1.18	1.066	1.323	1.143	1.16
Cob diameter	0	2.69	2.77	**2.93**	1.677
Cob length	0	11.46	8.43	**11.64**	6.37
Number of kernel row per ear	0	11.04	10.76	**11.35**	6.64
Number of kernel row	0	12.26	13.26	**13.825**	7.82
Number of kernel per ear	0	56.46	49.73	**102.37**	42
Silk receptivity	0	27.1	21.76	**35.95**	17.23
Shelling percentage	0	51.17	25.44	**50.89**	26.78
Thousand kernel weight (g)	0	285.32	260.467	271.825	164.76
Grain yield (kg ha^{-1})	0	537.68	293.4	**688.05**	316.04

Table 15: Distance among the different cluster centroid of maize under heat stress at NMRP, Rampur (2016).

	Cluster1	Cluster2	Cluster3	Cluster4
Cluster 1	0	614.977	398.599	750.397
Cluster 2		0	247.245	158.825
Cluster 3			0	399.583
Cluster4				0

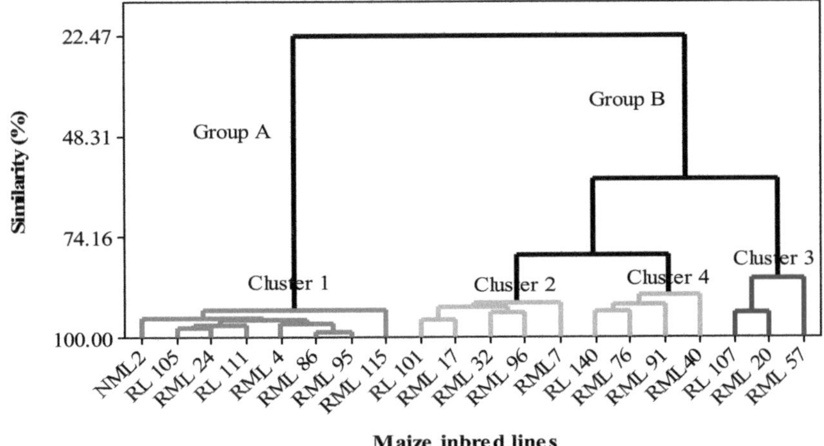

Figure4: Cluster analysis of the 20 maize inbred lines evaluated for agro-morphological traits under heat stress at NMRP, Rampur, Chitwan (2016).

Principal Component analysis

The PCA showed close resemblance with clustering and partitioned the total variance into 4 PCs having eigen value >1 explanining 85.9 % of total vaiation with eigen value between 10.935 to 1.293, among 20 genotypes of maize under heat stress,However, the remaning componnent contributed only 14.4 % towards total diversity for this set of maize genotypes.The first three pricipal component (PC1) which explained 79.2 % was associated mainly by anthesis silking interval (0.257), tassel blast (0.229), leaf firing (0.216) and leaf area index (0.120) with negative loading with grain yield, cob diameter (-0.295),cob length (-0.275),number of kernel ear^{-1}(-0.275),number of kernel row^{-1}(-0.297),number of kernel ear^{-1}(-0.298),shelling percentage(-0.276),silk receptivity (-0.295) and thousand kernel weight(-0.297) due to some heat susceptible genotypes in this cluster. Second principal (PC2) was responsible for about 13.4 % was mainly related to positively for leaf firing(0.239), physiological maturity (0.258).SPAD(0.145) and anthesis silking interval (0.106)with negatively for leaf area index(-0.273),tassel blast(-0.176),ear height(-0.538),ear per plant(-0.258),leaf senescence(-0.149) and grain yield(-0.008).PC3 contributed 8.3%with major positive contributor are leaf senescence (0.585) fallowed by leaf firing (0.218) whereas negatively associated with anthesis-silking interval (-0.104) leaf area index(-0.283),plant height(-0.147),ear height(-0.186), physiological maturity(-0.565) and grain yield(-0.135).PC4 accounted 6.8 % of the total variation was mainly negative association with, leaf area index (-0.489),leaf firing (-0.024),tassel blast (-0.139), leaf senescence(-0.397),SPAD(-0.201),shelling percentage (-0.058) and thousand kernel weight(-0.009) and positive association with anthesis-silking interval (0.154),grain yield (0.024),ear per plant(0.708),silk receptivity(0.030),plant height(0.708), number of kernel row ear^{-1}(0.026),number of kernel

29

ear^{-1}(0.087) and number of kernel row^{-1}(0.087).Thus positive relation with grain yield, anthesis silking interval, number of kernel row ear^{-1}, number of kernel ear^{-1} and number of kernel row^{-1},etc. and negative association with tassel blast, leaf firing, leaf sensencence, ear height, thousand kernel weight, shelling percentage lead to this principal component had variability and selection within this is importance for heat stress condition. The present research revealed that these genotype formed in four cluster in stress condition were the most suitable for cultivation under studied condition The finding PCA supported the result obtained by cluster analysis and PCA score plot was shown in figure 4.

Table 16: The first four principal components of traits used for cluster analysis and PCA and the eigen analysis of the correlation matrix at NMRP, Rampur (2016).

Variable	Principal	components		
	PC1	PC2	PC3	PC4
Eigen value	10.935	2.545	1.569	1.293
Proportion	0.578	0.133	0.083	0.068
Cumulative	0.578	0.71	0.792	0.860
Anthesis silking interval	0.257	0.106	-0.104	0.154
Leaf area index	0.120	-0.273	-0.283	-0.489
SPAD Chlorophyll	-0.200	0.145	-0.216	-0.201
Leaf firing	0.216	0.239	0.218	-0.024
Tassel blast	0.229	-0.176	0.074	-0.139
Plant height	0.03	-0.591	-0.147	0.039
Ear height	0.097	-0.538	-0.186	-0.035
Physiological maturity	0.103	0.258	-0.565	-0.009
Leaf senescence	-0.086	-0.149	0.585	-0.397
Ear per plant	0.025	-0.265	0.141	0.708
Cob diameter	-0.295	-0.031	0.046	-0.023
Cob length	-0.275	-0.032	0.001	-0.011
Number of kernel row per ear	-0.297	-0.039	0.019	0.026
Number of kernel row	-0.298	-0.003	-0.000	0.048
Number of kernel per ear	-0.275	-0.052	-0.192	0.087
Silk receptivity	-0.295	-0.02	-0.093	0.030
Shelling percentage	-0.281	-0.016	-0.008	-0.058
Thousand kernel weight	-0.297	-0.015	0.070	-0.009
Grain yield (kg ha^{-1})	-0.286	-0.008	-0.135	0.022

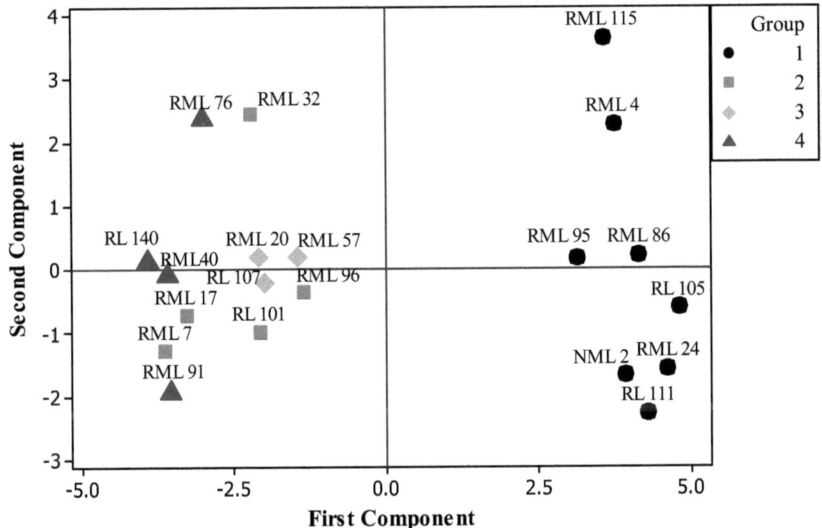

Fig 5: The score plot of first two components of maize inbred lines under heat stress

Conclusion

The genetic diversity was observed in inbred lines differences for grain yield and anthesis silking interval, SPAD reading and leaf senescence, tassel blast and leaf firing percentage, plant and ear height, leaf area index, ear per plant, cob length and diameter, number, kernel ear^{-1}, number of kernel row^{-1} and number kernel row ear^{-1}, silk receptivity, shelling percentage, thousand kernel weight under heat stress condition. Presence of additive gene effect along with positive correlation for grain yield with number of kernel ear^{-1}, silk receptivity, shelling percentage, and thousand kernel weight indicate direct selection and could be used as target traits to improve maize grain yield under heat stress condition but plant height and ear height exhibited non-significant correlated with grain yield hence may not be statistically considerable. But traits anthesis silking interval, tassel blast, leaf firing and leaf area index showed significant negative association with grain yield along with non-additive gene action suggested as hybrid plants have higher capacity to tolerate heat stress in field conditions than their parents and required further genotyping for development of thermo-tolerance maize inbred lines. SPAD reading, cob diameter and length, thousand kernel weight, silk receptivity, shelling percentage, number of kernel ear^{-1} with minimum tassel blast, leaf firing, shorter anthesis-silking interval were most yield determinative traits as revealed from correlation analysis and hence simultaneous selection for these trait might brining an improvement in maize grain yield under heat stress. Beside the correlation analysis inter se association also provide huge support on these traits from all other yield components. Path analysis using grain yield as dependent variable revealed that of thousand kernel weight, shelling percentage, number of kernel ear^{-1},silk receptivity exerted maximum positive direct effect on grain yield and these trait could be relied upon for selection of genotypes to improve grain yield. On contrary, some character via anthesis silking interval and tassel blast exerted positive direct effect on grain yield. However positive direct effect of effective theses trait were nullified by their negative effects through other components traits thousand kernel weight, shelling percentage, silk receptivity and number of kernel ear^{-1} which ultimately resulted in to highly significant negative correlation with grain yield. Hence indirect selection through other component characters with which these two traits exhibited negative indirect effects can be recommended so as to bring improvement in grain yield. Thus selection of genotypes having maximum thousand kernel weight, shelling percentage, silk receptivity and number of kernel ear^{-1} and minimum anthesis silking interval, leaf firing, tassel blast is pre-requisite for attaining improvement in grain yield under heat stress condition. SSI, STI, MP and GMP indices should help to improve heat tolerance in inbred lines. GMP and MP showed high positive correlations with grain yield in both stressed and normal environments should be more efficient in inbred line selection. Therefore, selection based on a combination of indices may provide a more useful criterion for improving heat tolerance and potential yield in different environment. UPGMA revealed that inbred lines formed four distinct clusters. The resistant lines and susceptible lines formed different clusters. The member of cluster 4 was found to be tolerance to heat stress where as members of cluster 1 were found most susceptible to heat stress. From this study inbred lines RL-140, RML-76, RML-91 and RML-40 were found most tolerant to heat stress as shown by lower reduction in grain yield and heat tolerance trait tassel blast, leaf firing and shorter anthesis siliking interval .Based on biplot analysis, lines RML-140 and RML-91 appeared as having high yielding and low stress susceptibility in both environments. Therefore, regarding the source of heat stress tolerance, we concluded that RML-140 and RML-91 should be a source of heat tolerance in crosses for hybrid production.

Acknowledgments

The authors have to mention Dr. Madhav Pandey, Associate Professor, Dr. Krishna Hari Dhakal, Mr. Raju Kharel, Department of Genetics and Plant Breeding, Agriculture and Forestry University (AFU) for providing suggestion and idea during the research. They would also like to thankful to National Maize Research Program Rampur (NMRP), Rampur,Chitwan for providing the genotype of maize inbred lines for research purpose. Also, this work has not been possible without the financial support of Nepal Agriculture Research and Development Fund (NARDF). Last but not the least, They would like to express my gratitude and love to my father, mother and whole family whose support sacrifices and selfless love has always been a driving force for my every endeavor.

References

Ali, F., Kanwal, N., Ahsan, M., Ali, Q., & Niazi, N. K. (2015). Crop improvement through conventional and non-conventional breeding approaches for grain yield and quality traits in *Zea mays* L. *Life Science Journal, 12*(4s).

Anderson, S. R., Lauer, M. J., Schoper, J. B., & Shibles, R. M. (2004). Pollination timing effects on kernel set and silk receptivity in four maize hybrids. *Crop Science, 44*(2), 464-473.

Asian Development Bank. (2009).The economics of climate change in Southeast Asia:In: A regional review. Retrieved 9 July 2018.

Becker, W. A. (1984). *Manual of quantitative genetics* (ed. 4). WA: Academic Enterprises.

Bello, O. B., Ige, S. A., Azeez, M. A., Afolabi, M. S., Abdulmaliq, S. Y., & Mahamood, J. (2012). Heritability and genetic advance for grain yield and its component characters in maize (*Zea mays* L.). *International Journal of Plant Research, 2*(5), 138-145.

Betrán, F. J., Beck, D., Bänziger, M., & Edmeades, G. O. (2003). Secondary traits in parental inbreds and hybrids under stress and non-stress environments in tropical maize. *Field Crops Research, 83*(1), 51-65.

Betrán, F. J., Beck, D., Bänziger, M., & Edmeades, G. O. (2003).Secondary traits in parental inbreds and hybrids under stress and non-stress environments in tropical maize. *Field Crops Research, 83*(1), 51-65.

Bolanos, J., & Edmeades, G. O. (1996).The importance of the anthesis-silking interval in breeding for drought tolerance in tropical maize. *Field Crops Research, 48*(1), 65-80.

Burton, G. W., & Devane, E. H. (1953). Estimating heritability in tall fescue (Festucaarundinacea) from replicated clonal material. *Agronomy Journal, 45*(10), 478-48.

Cairns, J. E., Sonder, K., Zaidi, P. H., Verhulst, N., Mahuku, G., Babu, R., and Rashid, Z. (2012). Maize Production in a Changing Climate: Impacts, Adaptation, and Mitigation Strategies. *Advances in agronomy, 114* (1).

Chen, J., Xu, W., Burke, J. J., & Xin, Z. (2010). Role of phosphatidic acid in high temperature tolerance in maize. *Crop science, 50*(6), 2506-2515.

Commuri, P. D., & Jones, R. J. (2001). High temperatures during endosperm cell division in maize. *Crop Science, 41*(4), 1122-1130.

Crafts-Brandner, S. J., & Salvucci, M. E. (2002). Sensitivity of photosynthesis in a C4 plant, maize, to heat stress. *Plant Physiology, 129*(4), 1773-1780.

Dupuis, I., & Dumas, C. (1990). Influence of temperature stress on in vitro fertilization and heat shock protein synthesis in maize (*Zea mays* L.) reproductive tissues. *Plant physiology, 94*(2), 665-670.

Edreira, J. R., Carpici, E. B., Sammarro, D., & Otegui, M. E. (2011). Heat stress effects around flowering on kernel set of temperate and tropical maize hybrids. *Field Crops Research, 123*(2), 62-73.

El-Badawy, M. E. M., & Mehasen, S. (2011). Multivariate analysis for yield and its components in maize under zinc and nitrogen fertilization levels. *Australian journal of basic and applied sciences, 5*(12), 3008-3015.

Farshadfar, E., & Sutka, J. (2002).Screening drought tolerance criteria in maize. *ActaAgronomicaHungarica, 50*(4), 411-416.

Fernandez, G. C. (1992, August). Effective selection criteria for assessing plant stress tolerance. In *Proceedings of the international symposium on adaptation of vegetables and other food crops in temperature and water stress* (pp. 257-270).

Gambín, B. L., Borrás, L., & Otegui, M. E. (2006). Source–sink relations and kernel weight differences in maize temperate hybrids. Field Crops Research, 95(2), 316-326.

Hussain, T., Khan, I. A., Malik, M. A., & Ali, Z. (2006).Breeding potential for high temperature tolerance in corn (*Zea mays* L.). *Pakistan Journal of Botany, 38*(4), 1185.

Hussain, T., Khan, I. A., Malik, M. A., & Ali, Z. (2006). Breeding potential for high temperature tolerance in corn (*Zea mays* L.). *Pakistan Journal of Botany, 38*(4),1185.

JawariaAzhar, J., Ramzan, J., & Ahmad, R. M. (2016). Reproductive response based diversity in genetically distant maize (*Zea mays* L.) germplasm under heat stress. *Journal of Agriculture & Basic Sciences*, 2518-4210.

Kamara, A. Y., Menkir, A., Badu-Apraku, B., & Ibikunle, O. (2003). Reproductive and stay-green trait responses of maize hybrids, improved open-pollinated varieties and farmers's local varieties to terminal drought stress. *Maydica, 48*(1), 29-38.

Kaur, R., Saxena, V. K., & Malhi, N. S. (2010).Combining ability for heat tolerance traits in spring maize (*Zea mays* L.). *Maydica, 55*(3), 195.

Khanna-Chopra, R., & Viswanathan, C. (1999). Evaluation of heat stress tolerance in irrigated environment of T. aestivum and related species. I. Stability in yield and yield components. *Euphytica, 106*(2), 169-180.

Khodarahmpour, Z., & Choukan, R. (2011). Genetic Variation of Maize (*Zea mays* L.) Inbred Lines in Heat Stress Condition. *Seed & Plant Improvement Journal, 27*(4), 539-554.

Khodarahmpour, Z., & Choukan, R. (2011). Genetic Variation of Maize (*Zea mays* L.) Inbred Lines in Heat Stress Condition. *Seed & Plant Improvment Journal, 27*(4), 539-554.

Krasensky, J., & Jonak, C. (2012). Drought, salt, and temperature stress-induced metabolic rearrangements and regulatory networks. *Journal of experimental botany, 63*(4), 1593-1608.

Krishnaji, J., Kuchanur, P. H., Zaidi, P. H., Ayyanagouda, P., Seetharam, K., Vinayan, M. T., & Arunkumar, B. (2017). Association and path analysis for grain yield and its attributing traits under heat stress condition in tropical maize (*Zea mays* L.). *Electronic Journal of Plant Breeding, 8*(1), 336-341.

Lobell, D. B., & Field, C. B. (2007). Global scale climate–crop yield relationships and the impacts of recent warming. *Environmental research letters, 2*(1), 014002.

MOAD, 2015/16.Statistical information on Nepalese agriculture.Agribusiness Promotion and statistics Division Singh Durbar, Kathmandu.Ministry of Government of Nepal.

Rafiq, C. M., Rafique, M., Hussain, A., & Altaf, M. (2010).Studies on heritability, correlation and path analysis in maize (*Zea mays* L.). *J. Agric. Res, 48*(1), 35-38.

Rafique, M., Hussain, A., Mahmood, T., Alvi, A. W., & Alvi, M. B. (2004). Heritability and interrelationships among grain yield and yield components inmaize (*Zea mays* L.). *International Journal of Agriculture and Biology (Pakistan).*

Robinson, H. F., Comstock, E., & Harvey, P. H. (1949). Estimation of heritability and degree of

Rosielle, A. A., & Hamblin, J. (1981).Theoretical aspects of selection for yield in stress and non-stress environment. *Crop science, 21*(6), 943-946.

Rowhani, P., Lobell, D. B., Linderman, M., & Ramankutty, N. (2011).Climate variability and crop production in Tanzania. *Agricultural & Forest Meteorology, 151*(4), 449-460.

Saxena, N. P., & O'Toole, J. C. (2002). Field Screening for Drought Tolerance in Crop Plants . International Crops Research Institute for the Semi-Arid Tropics. In: Rice Proceedings of an International Workshop on Field Screening for Drought Tolerance in Rice 11-14 Dec 2000.

Schlenker, W., & Roberts, M. J. (2009). Nonlinear temperature effects indicate severe damages to US crop yields under climate change. *Proceedings of the National Academy of sciences, 106*(37), 15594-15598.

Singh, B. D. (2001). Plant Breeding: Principles and Methods. Kalyani Publishers, New Delhi, India, Pages: 896.

Sivasubramanian, S., & Menon, M. (1973). Heterosis and inbreeding depression in rice. *Madras Agricultural Journal, 60,* 11- 39.

Steel, R. G. D., & Torrie, J. H. (1980). Principles and procedures of statistics, a biochemical approach. McGraw Hill, Inc. New York.

Steven J., Brandner, C., & Salvucci, M. (2002). Sensitivity of photosynthesis in C4 maize plant to heat stress. *Plant Physiol, 129,* 1773-1780.

Sumathi, P., Nirmalakumari, A., & Mohanraj, K. (2005). Genetic variability and traits interrelationship studies in industrially utilized oil rich CIMMYT lines of maize (*Zea mays* L). *Madras Agric. J, 92*(10-12), 612-617.

Van der Velde, M., Wriedt, G., & Bouraoui, F. (2010). Estimating irrigation use and effects on maize yield during the 2003 heatwave in France. *Agriculture, ecosystems & environment, 135*(1), 90-97.

Weaich, K., Bristow, K. L., & Cass, A. (1996). Modeling preemergent maize shoot growth: II. High temperature stress conditions. *Agronomy Journal, 88*(3), 98-403.

YOUR KNOWLEDGE HAS VALUE

- We will publish your bachelor's and master's thesis, essays and papers

- Your own eBook and book - sold worldwide in all relevant shops

- Earn money with each sale

Upload your text at www.GRIN.com and publish for free